915

Bre

Jellyfish
and Other Stingers

Concept and Product Development: Editorial Options, Inc.
Series Designer: Karen Donica
Book Author: Joseph K. Brennan

For information on other World Book
products, visit us at our Web site at
http://www.worldbook.com

For information on sales to schools and libraries
in the United States, call 1-800-975-3250.

For information on sales to schools and libraries
in Canada, call 1-800-837-5365.

World Book, Inc.
233 N. Michigan Avenue
Chicago, IL 60601

Library of Congress Cataloging-in-Publication Data

Jellyfish and other stingers.
      p. cm. -- (World Book's animals of the world)
    ISBN 0-7166-1221-6-- ISBN 0-7166-1211-9 (set)
      1. Jellyfishes--Juvenile literature. 2. Cnidaria--Juvenile literature. [1. Jellyfishes.
    2. Coelenterates.] I. World Book, Inc. II. Series.

    QL377.S4 J46 2001
    593.5'3--dc21   2001017524

Printed in Singapore

1 2 3 4 5 6 7 8 9 05 04 03 02 01

Jellyfish
and Other Stingers

Why don't I taste good with peanut butter?

World Book, Inc.
A Scott Fetzer Company
Chicago

Contents

What is so special about our friendship?

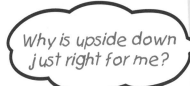

Why is upside down just right for me?

Why do people call me "Bud"?

What Is a Stinger?

The strange yet beautiful creature you see here is a jellyfish. Jellyfish are not fish—although they do spend their entire lives in water. Rather, these soft-bodied animals and their relatives are called cnidarians *(ny DAIR ee uhnz).* Cnidarians all have one thing in common: stinging cells. Some people call these creatures stingers.

This stinger is a compass jellyfish. Its body is shaped like an umbrella. There are also cnidarians that don't look at all like this compass jellyfish. Some have square shapes, while others look like flowers or plants.

In this book, you'll read about many kinds of jellyfish. You will also read about some other stinging cnidarians—hydras, sea anemones, and corals.

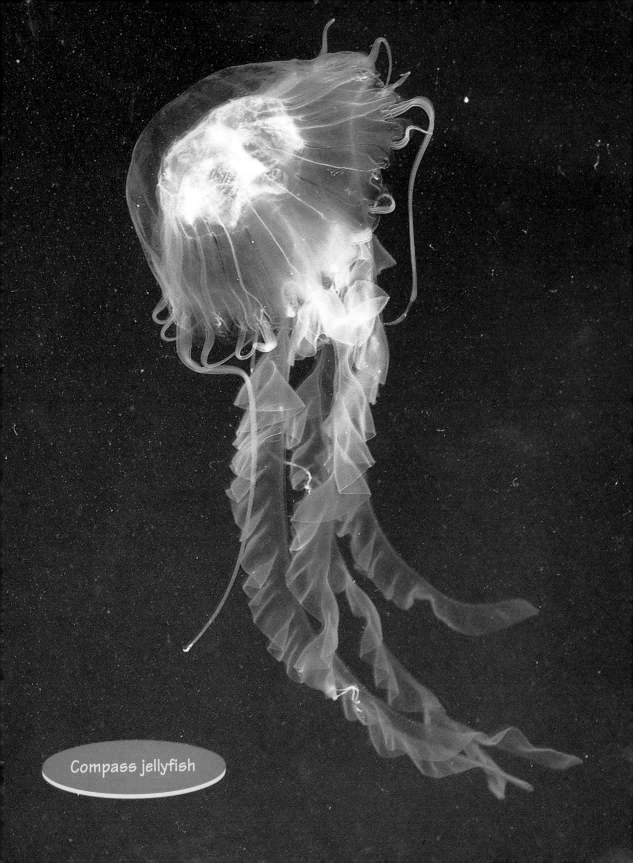

Compass jellyfish

Where in the World's Oceans Do Stingers Live?

Stingers live in all the oceans in the world. Some live in warm tropical waters. Others live in the icy waters of the Arctic. A few stingers even live in fresh water!

You may have seen jellyfish floating near the surface of the water. In fact, most jellyfish live in the upper waters of the ocean—the sunlit zone. At night, these jellyfish may sink down into deeper waters. Some stingers live their entire lives in deeper waters. Still others—such as sea anemones and corals—attach themselves to the ocean floor.

Jellyfish may swim or drift along with the ocean currents. Some swim alone. Others are found in huge groups, or shoals *(SHOHLZ)*. Wherever they are, these unusual creatures are always moving.

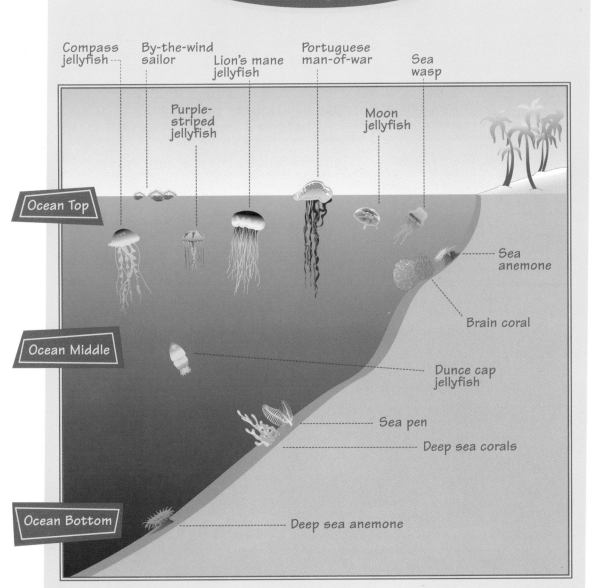

Ocean Depths of Jellyfish

Compass jellyfish

By-the-wind sailor

Lion's mane jellyfish

Purple-striped jellyfish

Portuguese man-of-war

Moon jellyfish

Sea wasp

Ocean Top

Sea anemone

Brain coral

Ocean Middle

Dunce cap jellyfish

Sea pen

Deep sea corals

Ocean Bottom

Deep sea anemone

Are Jellyfish Made of Jelly?

If you are thinking of the jelly you eat with peanut butter, the answer is no. That jelly is fruit and fruit juice mixed with a thickener.

But jellyfish are made of another kind of jelly. Jellyfish jelly is an amazing blend of salt, protein, and lots of water. In fact, most jellyfish are over 95 percent water.

A jellyfish has a very simple body. Two layers of cells make up the body walls. A thick layer of jellyfish jelly between these walls helps give the jellyfish its shape.

Jelly is a perfect material for the body of an animal that lives in water. It helps support the jellyfish in the water. It helps jellyfish that live near the surface to float. And it keeps deep-sea jellyfish from being crushed by the pressure of the water.

Purple-striped jellyfish

What Are the Parts of a Jellyfish?

Jellyfish have no gills or lungs. They have no hearts or brains. In fact, jellyfish have no bones or skeletons to give them shape or to support their bodies. The jelly does that!

A jellyfish's body looks a lot like a bell or an umbrella. Scientists call this body style a medusa *(muh DOO suh)*. A jellyfish's mouth is a small opening on the underside of the bell. It is the only opening leading into the jellyfish and the only opening leading out.

As you can see from the diagram, a jellyfish has many tentacles. Some hang around the bell of the jellyfish. These hold the stinging cells. Other tentacles hang around the mouth. These are called oral arms. They may or may not have stinging cells. The oral arms pass the food into the jellyfish's mouth. The food then goes into the animal's stomach.

Diagram of Jellyfish

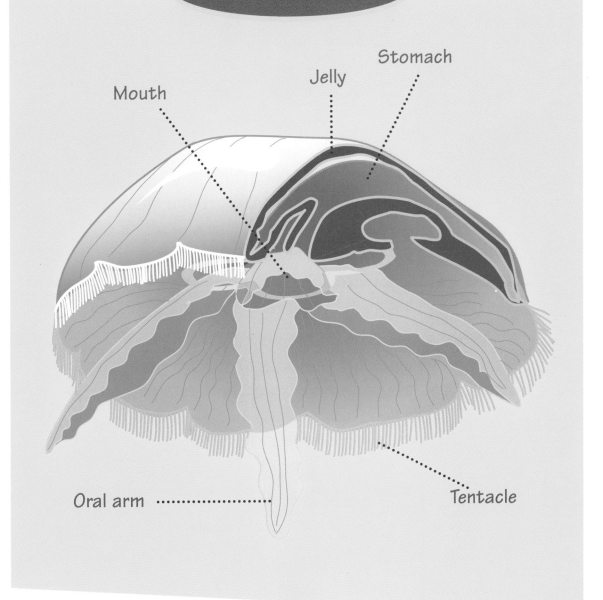

Stomach

Jelly

Mouth

Oral arm

Tentacle

13

How Do Jellyfish Sense Their World?

Humans learn about their world by using their five senses. But most jellyfish don't have eyes, ears, a nose, a tongue, or hands. These jellyfish depend on simple sense cells.

Look at this moon jellyfish. You can probably find the oral arms and the tiny tentacles around the edge of its bell. But what might be harder to see are the special sense cells. Look carefully. Do you see ridges along the edge of the bell? That is where the jellyfish's sense cells are located.

These sense cells perform different jobs. One kind acts as eyes and senses light. Sensing light helps the jellyfish know if it is upside down in the water. Another kind acts as a nose and senses chemicals in the water. A third kind helps a jellyfish keep its balance in the water.

Moon jellyfish

What Do Jellyfish Eat?

Jellyfish feed on fish and other sea animals. They also eat very small animals. Some of these animals are so small that they are microscopic. That means you need a microscope in order to see them. Scientists call this collection of small animals zooplankton *(ZOH uh PLANG tuhn).* Zooplankton include tiny shrimp, eggs of sea animals, baby sea animals, and even other stingers.

How small are zooplankton? Take a look at the picture. It shows zooplankton under a microscope. And these are just some of the animals you might find in a single drop of water! Now, you may be thinking that animals this small are not much of a meal. But to a jellyfish, thousands and thousands of zooplankton do add up.

Zooplankton in a drop of water

How Do Jellyfish Find Food?

Jellyfish do not hunt for food as many animals do. Instead, they take in food as they swim through the water. Jellyfish use their oral arms to sweep zooplankton into their mouths.

Jellyfish also use their tentacles to capture larger prey. When a fish brushes against a jellyfish's tentacles, stinging cells stun or kill the fish. Then the tentacles bring the fish into the jellyfish's mouth.

The jellyfish you see here is a sea nettle. Can you tell what it's eating for dinner?

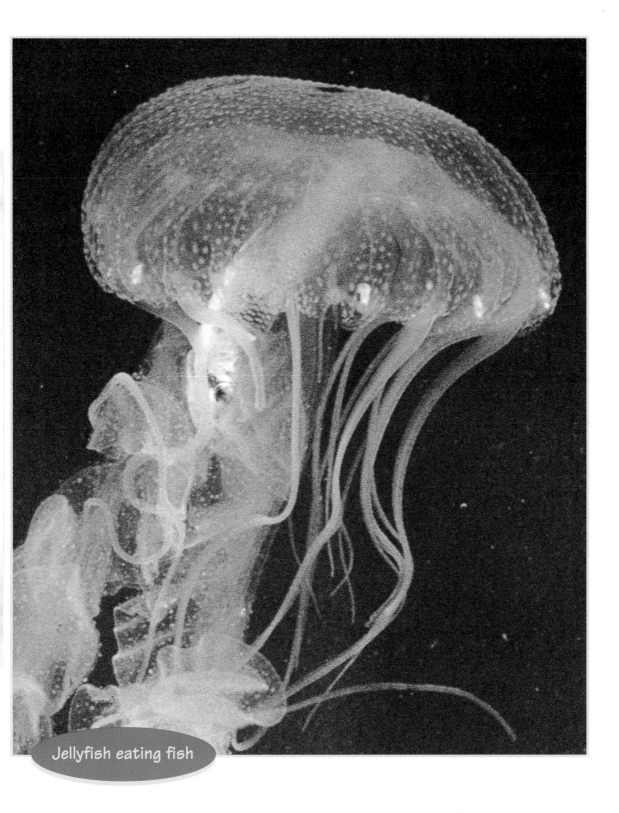

Jellyfish eating fish

Why Do Jellyfish Sting?

Jellyfish sting in order to catch their prey. They also sting to keep other animals from eating them.

Jellyfish tentacles and oral arms are covered with thousands of stinging cells. Inside each stinging cell is a hollow tube. This tube looks like a coiled thread. When an animal touches the stinging cell, the tube fires out to sting the prey.

All cnidarians have stinging cells. But the cells may work in different ways to deliver their stings. Some stings pierce an animal's skin. These stings release a poison. The poison paralyzes the prey so that it cannot move. Other stings are very long. They wrap around the prey and trap it.

A jellyfish's sting is a good defense. But it does not protect a jellyfish from all sea animals. Some animals that eat jellyfish aren't affected by the stings. They include sea slugs, ocean sunfish, and sea turtles.

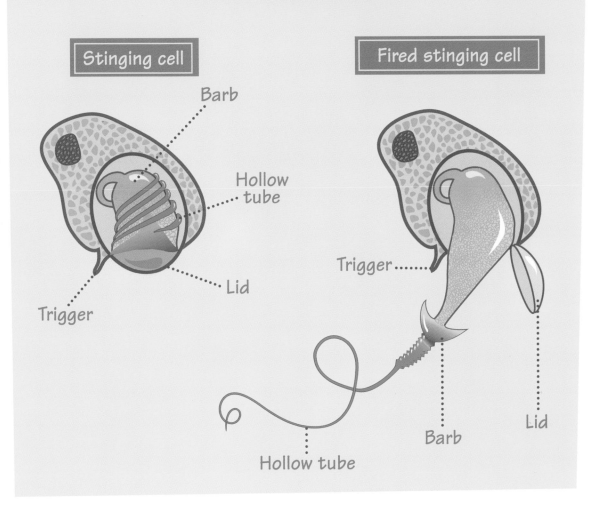

Diagram of Stinging Cell

Stinging cell

Barb

Hollow tube

Lid

Trigger

Fired stinging cell

Trigger

Barb

Lid

Hollow tube

21

Why Do Some Jellyfish Light Up?

Some jellyfish, such as sea nettles, make their own light. They glow or give off flashes of light as fireflies do. Some jellyfish use this light to attract prey. But most jellyfish use it as a defense against predators.

How do its lights help a jellyfish? A jellyfish may light up to surprise a predator or to frighten it away. Lighted up, a small jellyfish with long tentacles suddenly looks like a large animal.

One jellyfish drops its glowing tentacles when fleeing from a predator. The attacking animal swims after the falling tentacles. This gives the jellyfish time to get away.

Jellyfish may light up for other reasons, too. To find out what those reasons are, scientists dive down in ships called submersibles *(sub MER suh buhlz)*. The ships have windows that let scientists study the jellyfish deep in the ocean where they live.

Sea nettle

How Do Jellyfish Get Around?

Jellyfish swim to get around. For a jellyfish, this means opening and closing its bell—much as you open and close an umbrella. A jellyfish has a ring of muscle around its bell. When a jellyfish tightens this muscle, its bell closes. This pushes water inside the jellyfish out, shooting the jellyfish forward. As the muscle relaxes, water refills the bell.

Look at the swimming jellyfish in the picture. The jellyfish on the right is tightening its bell. It is shooting through the water.

Though jellyfish can swim, ocean currents often control where they go. Jellyfish float, drift, and sink with the ocean currents.

Ocean currents help jellyfish move around. But currents can also cause a lot of damage. Jellyfish are very fragile. Strong currents during storms can tear a jellyfish's bell or break off its tentacles. Currents can also leave a jellyfish stranded on a beach.

24

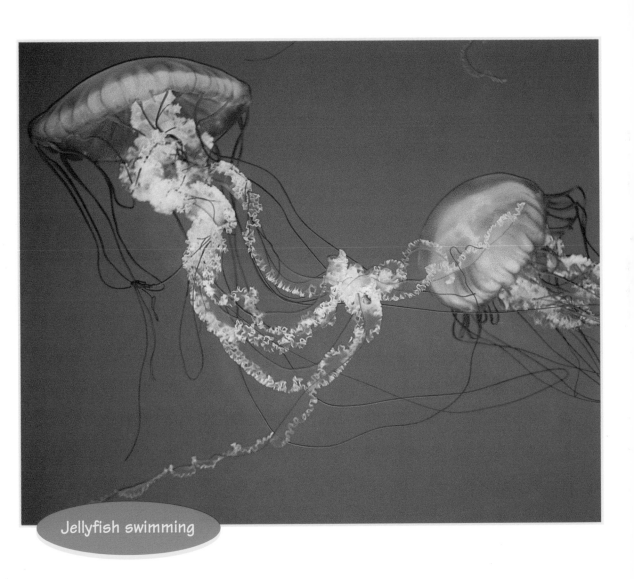

Jellyfish swimming

25

How Do Jellyfish Grow?

Young jellyfish look very different from their parents. In fact, a young jellyfish goes through several stages before it becomes an adult. Other animals go through life cycles, too. Think about the way in which a caterpillar becomes a butterfly or a tadpole becomes a frog.

Most large jellyfish start out as eggs. Look at the diagram to see what happens to a jellyfish egg. You'll see that a jellyfish egg develops into a larva *(LAR vuh).* The larva drifts along in the ocean currents until it settles and attaches itself to the ocean bottom.

Next, the larva grows into a polyp *(PAHL ihp).* As the polyp develops, its stem begins to look like a stack of tiny saucers. One by one, the "saucers" break off, and each one develops into a tiny medusa. In this way, a single larva produces many jellyfish.

26

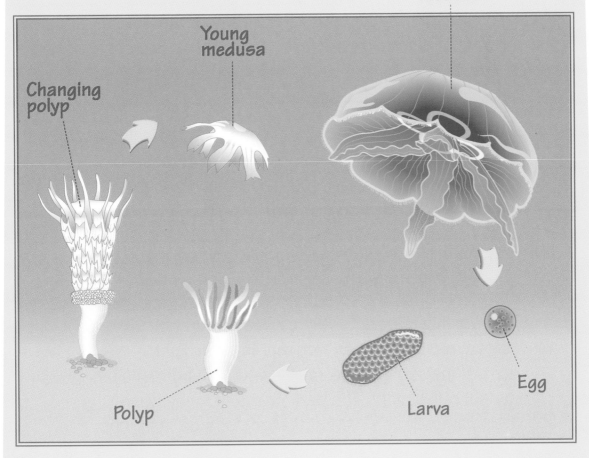

Life Cycle of a Jellyfish

Adult medusa

Young medusa

Changing polyp

Polyp

Larva

Egg

Why Are Some Jellyfish So Colorful?

Many jellyfish that live deep in the ocean are red—like this dunce cap jellyfish. Others are purple, brown, or black. These colors help jellyfish in two ways. They help hide what a jellyfish has eaten. And they make it hard for predators to see the jellyfish.

In the deep ocean, being able to hide a meal can be very important. The reason is that many deep-sea animals use light—much the same way jellyfish do. Imagine what would happen if a clear jellyfish ate a glowing fish. That's right! The fish would show through the jellyfish's bell. But a colored bell hides the meal, so predators cannot see the fish.

Very little sunlight reaches deep down into the ocean. A clear jellyfish would reflect this light, almost the same way a mirror would. But dark colors do not reflect light. The dark colors of the jellyfish help camouflage *(KAM uh flahzh),* or hide, it from its predators.

Dunce cap jellyfish

Which Is the Largest Jellyfish?

The lion's mane is the largest kind of jellyfish. Its bell can grow to widths of more than 7 feet (2.1 meters). And its long tentacles can extend more than 130 feet (40 meters). This jellyfish can be found in the cold waters of the Atlantic and Pacific oceans. The largest lion's manes are found in the Arctic Ocean.

The lion's mane jellyfish is named for its shaggy oral arms and hairlike tentacles. Its stings are toxic, or poisonous. To catch prey, this jellyfish sinks down in the water. It spreads its tentacles out around the prey. The tentacles act like a huge net to catch the prey.

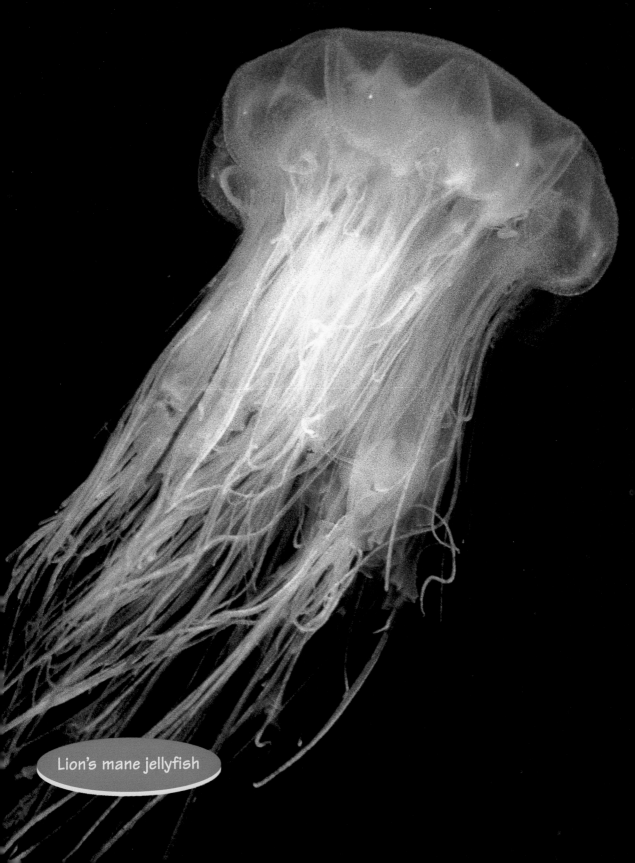

Lion's mane jellyfish

Which Jellyfish Spends Its Life Upside Down?

Most jellyfish swim through the water with their tentacles and oral arms hanging below them. But not Cassiopeia *(KAS ee uh PEE uh),* the upside-down jellyfish! This stinger spends most of its life with its tentacles floating above it.

Cassiopeia is a bottom-feeder. It lives in the shallow water of swamps near the seacoast. It sinks under the water and uses its bell as a suction cup to hold onto the bottom. Then Cassiopeia waves its oral arms to catch passing zooplankton.

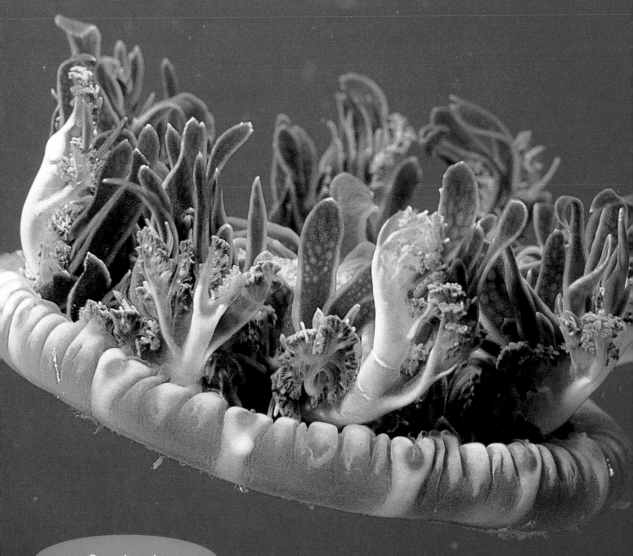

Cassiopeia

Which Jellyfish Has the Most Powerful Sting?

Sea wasps have powerful stings that can cause severe burns. But the Australian sea wasp is the most feared jellyfish of all. Its poison is deadlier than any snake venom. And it can kill large animals, including humans, in as little as three minutes.

The Australian sea wasp may grow as large as a basketball. And its threadlike tentacles can be as long as 15 feet (4.6 meters). But the bell is clear and very hard to see in the water. Australian sea wasps pose a special threat to swimmers. The reason is that these jellyfish often feed close to the shore where people swim.

Even Australian sea wasps, however, have enemies. Large sea turtles feed on them. The sea turtle's tough skin and stomach may protect it from the jellyfish's sting. But scientists aren't really sure how the sea turtle can swallow the jellyfish without being poisoned.

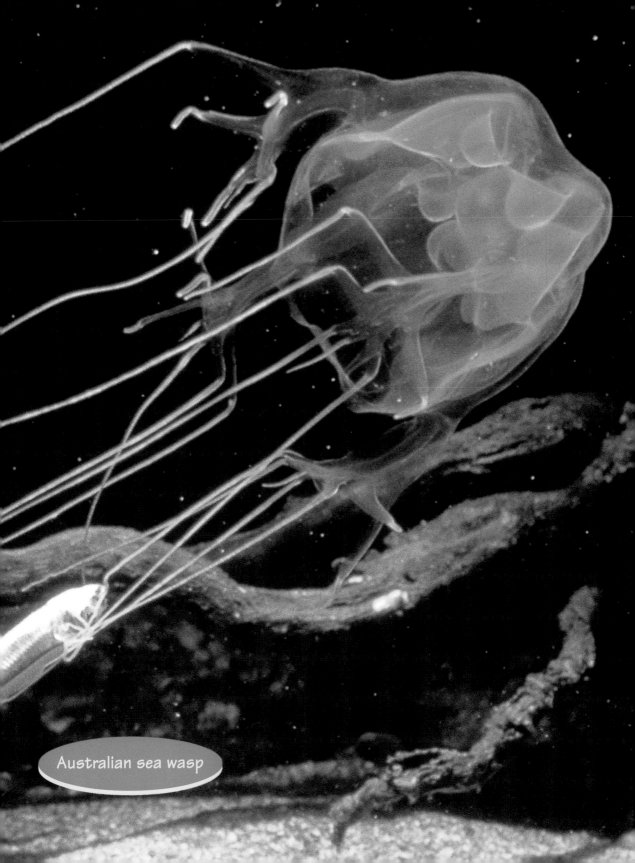

Australian sea wasp

Can Sea Wasps See in the Sea?

Sea wasps are not like most other jellyfish. Sea wasps are the only stingers that have eyes. But scientists aren't sure what sea wasps can see. Like other stingers, sea wasps don't have brains. So, it is hard to figure out how these stingers use their eyes. But they probably use them to find prey and keep away from their predators.

A sea wasp has four eyes—one on each of its four sides. In this picture, one of the eyes looks like a white dot in the middle of the sea wasp's medusa.

Sea wasps are different from most jellyfish in other ways, too. Sea wasps don't have oral arms. Instead, they have four groups of tentacles. The tentacles are found at each corner of their box-shaped medusa. Because of their unusual shape, these stingers are also known as box jellyfish.

Australian sea wasp

What Is a Portuguese Man-of-War?

The Portuguese (*PAWR chuh GEEZ*) man-of-war looks like a jellyfish. But it's not a jellyfish. This stinger is really a colony, or group, of hundreds of polyps living and working together.

The main part of the Portuguese man-of-war is a gas-filled float. The float's job is to keep the colony on the surface of the water. The float was also the first member of the colony. All the other polyps budded, or grew, from the float.

These other polyps hang from the float. Some are tentacles that sting and capture prey. Others are feeding polyps. They digest food and pass it along to the rest of the colony. Other polyps are in charge of reproduction.

A Portuguese man-of-war may have a float up to 10 inches (25 centimeters) long. It may have tentacles 30 feet (9 meters) long. Its sting is very powerful, but not so deadly as that of the Australian sea wasp.

Portuguese man-of-war

Which Stinger Has Its Own Sail?

There is one stinger that looks as if it has its own sail. It is called the by-the-wind sailor. Some scientists think that this stinger is really a colony of polyps—much as the Portuguese man-of-war is. Others believe it is just one big polyp.

By-the-wind sailors can't swim. Instead, they are carried along by the wind. After several days of strong winds, swarms of these jellies may stretch out for more than 1/2 mile (0.8 kilometer).

Animals that feed on by-the-wind sailors follow the swarms. These predators include sea snails, sea slugs, and the ocean sunfish mola mola.

By-the-wind sailors

What Looks Like a Jellyfish but Isn't?

The sea creature you see here is a kind of comb jelly. It's called a sea gooseberry. Like jellyfish, a comb jelly has two layers of cells that form its body wall. It also has a jellylike middle layer. But a comb jelly does not have stinging cells. So it is not a cnidarian.

Comb jellies do need to eat, however. How do they catch their prey if they can't sting? Many, like the sea gooseberry, have sticky cells on their tentacles. The comb jellies use these sticky cells to catch prey.

Comb jellies are named for the rows of comblike hairs that run along their bodies. These tiny hairs beat the water to push the comb jellies along. But comb jellies are poor swimmers. They depend on ocean currents to carry them along.

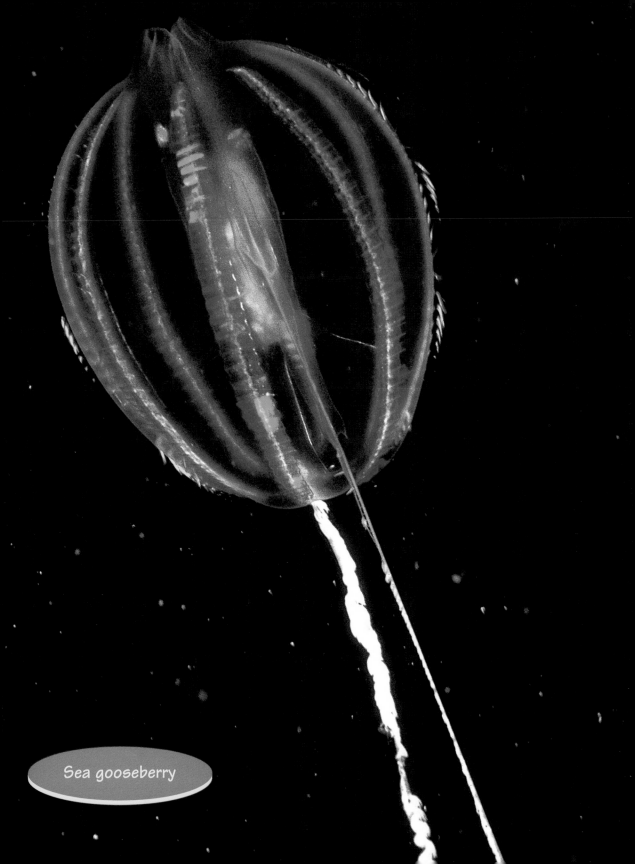

Sea gooseberry

Which Stingers Live in Fresh Water?

Most hydras *(HY druhz)* spend their whole lives as tube-shaped polyps. Still, they have a lot in common with jellyfish. A hydra has two cell walls with a thin layer of jelly in between. It has a mouth, tentacles, and lots of stingers.

Hydras live in freshwater ponds. They are *very* small—and look like bits of string with frayed ends. A typical hydra is only 1/4 to 1/2 inch (6 to 13 millimeters) long. The hydras you see in the photo are magnified, or shown larger than they really are.

To grow, a hydra attaches itself to the bottom of a pond or to a water plant. But it can and does move. A hydra can slide along on its cuplike foot. It can somersault or pull itself along by its tentacles. It can also let go and float free on an air bubble.

Hydras

What's So Special About Hydras?

Hydras can do something that is very rare among animals. They can regrow missing body parts! A hydra that has lost a tentacle can grow a new one. What's more, a hydra that has been cut in half can grow into two new hydras. And, would you believe this? If a hydra is cut into 50 pieces, it can grow into 50 new hydras!

Hydras usually reproduce by budding. A new hydra first appears as a small bud on the parent hydra's stem. Eventually, it breaks off and forms a new hydra.

Budding hydra

Which Polyp Looks Like a Flower?

A sea anemone *(uh NEHM uh nee)* looks a lot like a flower. But of course, it isn't. A sea anemone is a stinger that lives almost its whole life as a polyp. Some sea anemones also go through a larva stage.

The body of a sea anemone is a hollow column. The mouth and the tentacles are at the top. Like most stingers, a sea anemone uses its tentacles to capture prey and push it into its mouth.

A few sea anemones move around much in the same way that hydras do. Some creep along on the suction foot. Others somersault. And still others swim by flexing their bodies.

If a sea anemone is threatened, it pulls its tentacles into its body. This makes the animal look like a colorful ball.

Sea anemones

Why Is That Sea Anemone Riding a Crab?

Some sea anemones have special relationships with other animals. The relationships are special because each animal helps the other in some way. This sea anemone, for example, may spend much of its life clinging to the shell of the hermit crab. Why? The hermit crab helps the sea anemone catch more food by moving it from place to place. In turn, the sea anemone's tentacles help protect the hermit crab from its predators.

Some fish also have special relationships with sea anemones. One kind of fish, the clown fish, lives among the sea anemone's stinging tentacles. That keeps it safe from its predators. In turn, the clown fish chases away predators that might want to eat the sea anemone. The clown fish may also help keep the sea anemone clean.

Crab with sea anemone

Which Stinger Is a Master Builder?

A colony of tiny coral polyps, no bigger than pencil tips, built the coral structure you see here. This coral is named for what it looks like; it is brain coral. Brain coral is one of the many kinds of corals you can see in a shallow-water reef.

Coral polyps build structures by doing what other stingers can't do. First, they take in calcium *(KAL see uhm)* from the water. Then, they turn the calcium into limestone. Last, they deposit the limestone around their bodies. These stony "shelters" are safe places for coral polyps to hide from enemies. The polyps reach out with their tentacles and catch zooplankton.

When coral polyps die, they leave behind the shelters, or skeletons. New polyps build and add on to the skeleton. But it takes a long time—hundreds or even thousands of years—for a coral to grow as large as this brain coral.

Brain coral with diver

Who Builds a Skeleton on the Inside?

Some coral polyps build skeletons outside their bodies. They are stony corals. Others build skeletons inside their bodies. They are soft corals. Soft corals are formed by polyps that grow skeletons from protein or from calcium spikes. The amazing thing about these skeletons is that they are formed inside the bodies of the polyps!

The sea fan you see here is a soft coral. It has hundreds of tiny polyps living on its wide, lacy branches. The polyps are so small that their tentacles create a fuzzy fur all over the fan.

You may have seen jewelry made from coral. Artists carve the brightly colored skeletons into beads and other shapes. They polish the coral until it shines.

Sea fan

How Are Many Corals Named?

Many corals have been named for the objects they look like. Lettuce-leaf coral looks like the makings of a salad. Boulder coral looks like huge rocks. Staghorn coral resembles the antlers of a deer. Star coral has tiny star shapes covering it.

Many other corals also have such names. But some of these names are a little old-fashioned. The coral you see here, for example, is a sea pen. It probably doesn't look like a pen you might use. But it does look like a quill, or feather pen, that people used long ago.

Sea pen

What Is a Coral Reef Community?

Many animals and plants live in and around a coral reef. A coral reef is home to coral polyps and other stingers. It is home to fish and crabs. It is also home to algae *(AL jee)* and seaweed. Algae are tiny, plantlike things. All these plants and animals form a coral reef community.

Plants and animals in a community need one another. Here is something to think about: Coral reefs grow in warm, shallow waters, where sunlight can filter down. A coral reef is a good place for plants to grow. Fish that eat plants come to the reef. Fish that eat those fish also come to the reef. They live in the reef and lay their eggs here. The eggs and fish that hatch from them provide food for the coral polyps.

Look at this coral reef community. How many different sea animals can you find in the picture?

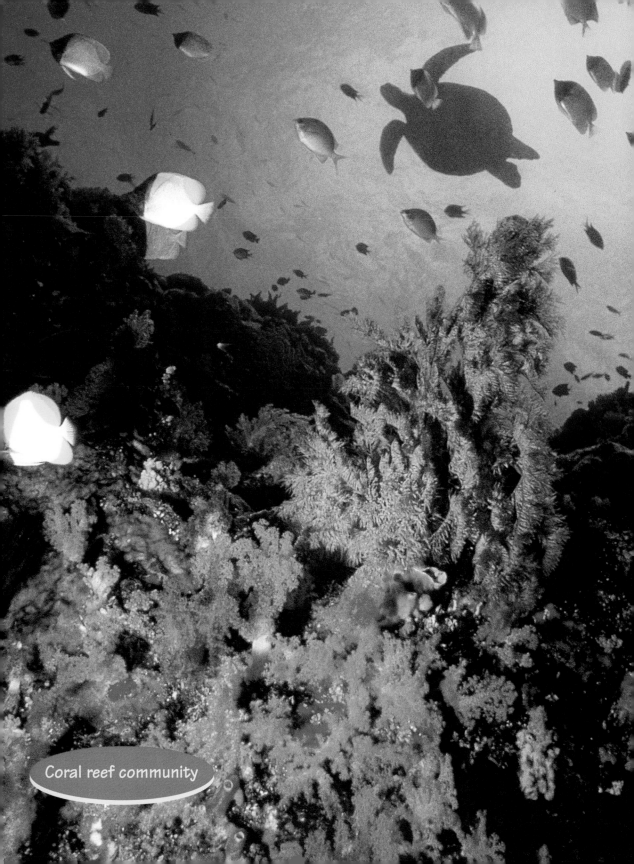

Coral reef community

Are Stingers in Danger?

Changes to the world's oceans do threaten some stingers. Perhaps the stingers most in danger are those in coral reefs. Most healthy coral polyps have algae living inside them. Like plants, algae store energy from sunlight. Coral polyps need this energy in order to survive.

In many coral reefs, the algae are dying. If the algae die, the coral polyps may die, too. Scientists think that pollution, disease, or water that is too warm may be the cause.

Activities like boating and diving also hurt a coral reef. Boaters may run into a reef or drop their anchors on the coral. Divers may kick or brush up against the delicate polyps. This type of damage can take years to heal.

Scientists are working to save the coral. Some reefs are being set aside as parks. And research is being done to help stop the coral from dying.

Coral reef

Stinger Fun Facts

→ A stinging cell can fire over 60 times faster than the blink of a human eye.

→ The Great Barrier Reef is the largest coral structure in the world. It extends about 1,250 miles (2,010 kilometers) along the northeast coast of Australia.

→ Many different crabs, fish, and shrimp hitch rides inside the bells of jellyfish.

→ The stinging cells of jellyfish can still fire after the jellyfish is dead.

→ When a clown fish is taken away from its sea anemone, it will swim right back to the sea anemone.

→ Sea slugs eat floating cnidarians without causing the stingers to fire their stings. The sea slugs use the stings to fend off their own predators.

→ Sea whips, sea fans, and sea pens don't like sunlight. These polyps open up only at night or on cloudy days.

Glossary

algae Plantlike bacteria.

bell The body of a jellyfish.

bud To grow off an animal until developing into a separate organism.

calcium A chemical found in milk, bones, and shells.

camouflage To change in appearance in order to hide.

cell The smallest unit of a plant or animal able to perform a job.

cnidarian A soft-bodied water animal that has tentacles with stinging cells.

colony A group of animals living and working together.

coral The skeleton formed by tiny sea animals massed together.

current A path of moving water.

larvae The newly hatched and wormlike offspring of jellyfish.

medusa The free-swimming adult stage of a jellyfish.

microscopic Too small to be seen without a microscope.

polyp A small animal with a tube-shaped body and a mouth surrounded by tentacles.

predator An animal that lives by hunting and killing other animals for food.

prey Any animal that is hunted for food by another animal.

oral arm A body part that grasps food or passes food into the mouth of a jellyfish.

reef A mass of coral that rises close to the surface of the water.

reproduction The way by which animals produce offspring.

shoal A large group of marine animals.

submersible A small ship that can operate underwater.

tentacle A narrow, flexible body part that certain animals use for feeling, grasping, or feeding.

toxic Very poisonous.

zooplankton Very small plants and animals that float in seas and lakes.

63

Index

(**Boldface** indicates a photo, map, or illustration.)

Picture Acknowledgments: Front & Back Cover: © Frieder Sauer, Bruce Coleman Collection; © R.N. Mariscal, Bruce Coleman Inc.; © Andrew J. Martinez, Photo Researchers; © Tom & Therisa Stack, Tom Stack & Associates; © F. Stuart Westmorland, Photo Researchers.

© William H. Amos, Bruce Coleman Inc. 23; © Tom Branch, Photo Researchers 45; © Ben Cropp, www.norbertwu.com 35, 37; © Dave B. Fleetham, Tom Stack & Associates 39; © Jeff Foott, Bruce Coleman Inc. 57; © Gilbert S. Grant, Photo Researchers 15; © Robert Hermes, Photo Researchers 19; © Byron Jorjorian, www.bjphoto.com 25; © Dwight R. Kuhn, 5, 47; © R.N. Mariscal, Bruce Coleman Inc. 4, 51; © Andrew J. Martinez, Photo Researchers 5, 33, 43; © Chris McLaughlin, Animals Animals 55; © Flip Nicklin, Minden Pictures 41; © Gregory Ochocki, Photo Researchers 11; © Peter Parks, www.norbertwu.com 17; © Frieder Sauer, Bruce Coleman Collection 3, 7; © Tom & Therisa Stack, Tom Stack & Associates 53; © F. Stuart Westmorland, Photo Researchers 49; © Norbert Wu, www.norbertwu.com 29, 31, 59, 61.
Illustrations: WORLD BOOK illustration by Michael DiGiorgio 9, 13, 21, 27; WORLD BOOK illustration by Karen Donica 62.

Stinger Classification

Scientists classify animals by placing them into groups. The animal kingdom is a group that contains all the world's animals. Phylum, class, order, and family are smaller groups. Each phylum contains many classes. A class contains orders, and a family contains individual species. Each species also has its own scientific name. Here is how the animals in this book fit in to this system.

Stingers (Phylum Cnidaria)

Anemones and corals (Class Anthozoa)

Anemones (Order Actiniaria)

Brain, leaf-lettuce, staghorn, and other stony corals (Order Scleractinia)

Sea fans, sea pens, and other soft corals (Orders Gorgonacea and Pennatulacea)

Hydras and their relatives (Class Hydrozoa)

Brown hydra . . . . *Hydra oligactis*
By-the-wind sailor . . . . *Velella velella*
Portuguese man-of-war . . . . *Physalia physalis*

Jellyfishes (Class Scyphozoa)

Cassiopeia . . . . *Cassiopeia xamachana*
Compass jellyfish . . . . *Chrysaora hyoscella*
Dunce cap jellyfish . . . . *Periphylla periphylla*
Lion's mane jellyfish . . . . *Cyanea capillata*
Moon jellyfish . . . . *Aurelia aurita*
Purple-striped jellyfish. . . . . *Pelagia colorata*
Sea nettle. . . . . *Chyrsaora quinquecirrha*

Sea wasps and box jellyfishes (Class Cubozoa)

Australian sea wasp. . . . . *Chironex fleckerii*

Comb jellies (Phylum Ctenophora)

Sea gooseberry . . . . *Pleurobrachia pileus*